Become a Math Beast!
For additional books,
printables, and more, visit

BeastAcademy.com

This is Guide 4D in a four-book series for fourth grade:

Contents:

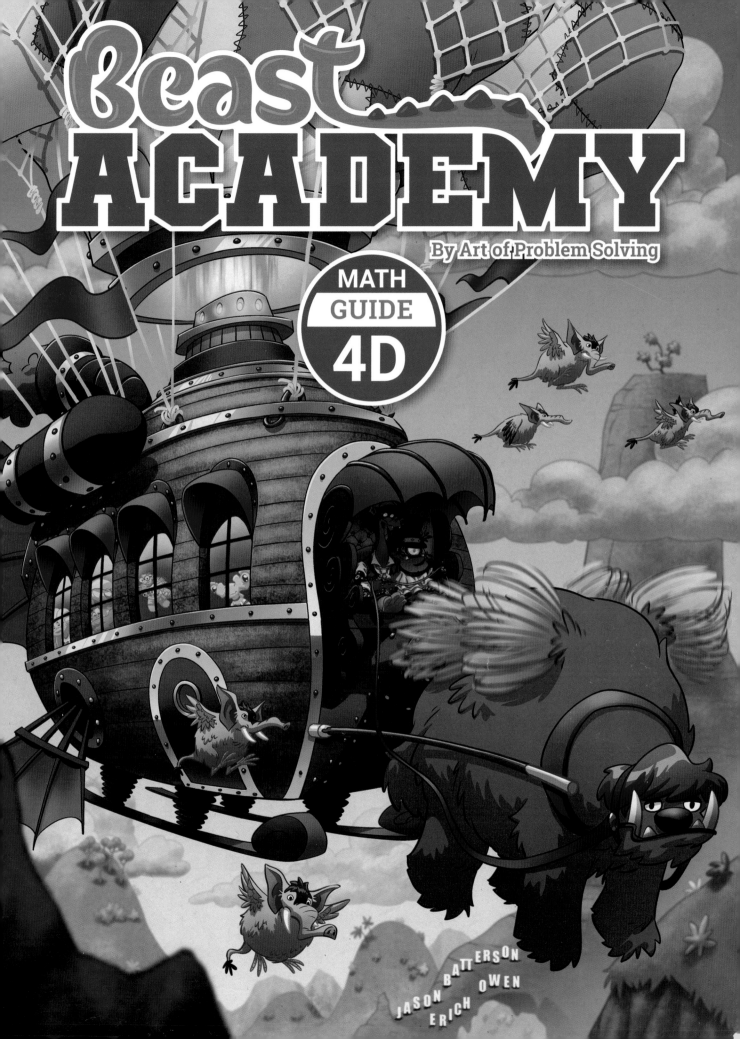

Published by: AoPS Incorporated
                   10865 Rancho Bernardo Rd Ste 100
                   San Diego, CA 92127-2102
                   info@BeastAcademy.com

ISBN: 978-1-934124-56-7

Written by Jason Batterson and Kyle Guillet
Illustrated by Erich Owen
Additional Illustrations by Paul Cox
Colored by Greta Selman

Visit the Beast Academy website at www.BeastAcademy.com.
Visit the Art of Problem Solving website at www.artofproblemsolving.com.
Printed in the United States of America.
2018 Printing.

Grok
Math lab
should probably hire a bodyguard
constantly captured by

calamitous clod
Almost always alliterates
Probably good at juggling

Ms. Q.
Math Teacher
May someday turn into a really big butterfly

Kraken
pirate shop Teacher

sergeant Rote
gym teacher
VR-960A Hoversphere
is the same size as a regulation basketball

Rosencrantz and Guildenstern
custodian(s?)
Have the same birthday!
when one eats, does the other feel full?

Fiona
math team coach
fastest 2-legged sprinter in her grade at B.A.

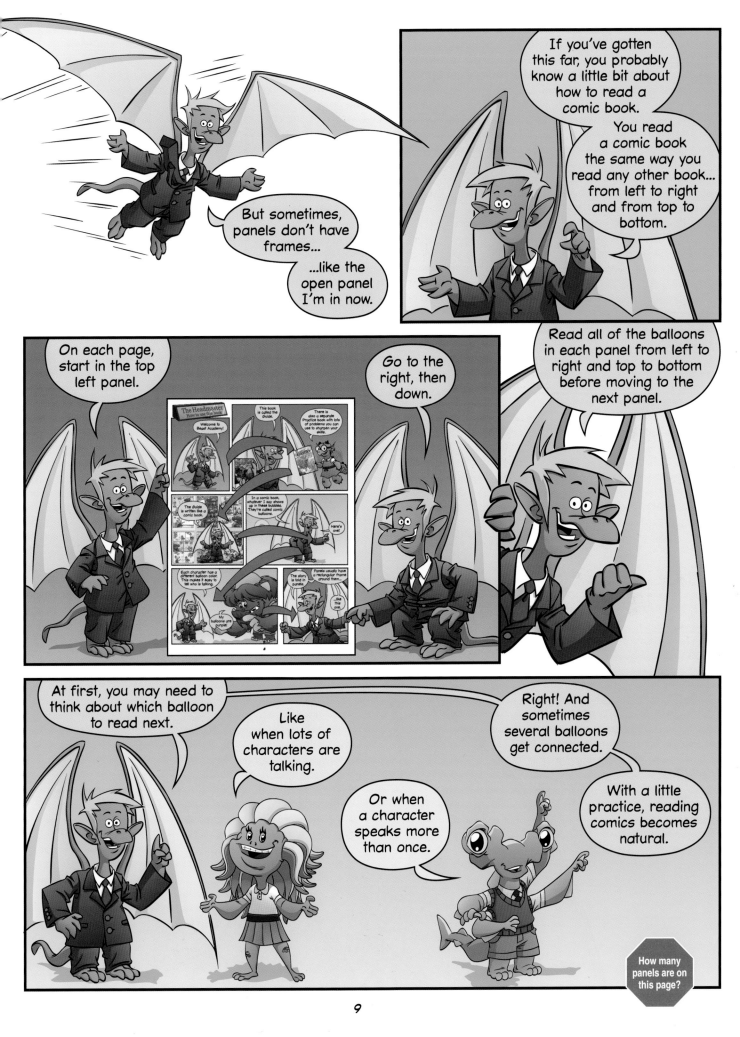

Contents: Chapter 10

See page 6 in the Practice book for a recommended reading/practice sequence for Chapter 10.

Chapter 10:
Fractions

There's no better workout than a 15-kilometer trek on cross-country skis.

15 kilometers!?

Yep. We're going all the way around Wampa Reservoir.

Hope we don't see any wampas!

We're already $\frac{1}{3}$ of the way around!

What's $\frac{1}{3}$ of 15?

If you split 15 into three equal parts, each part is $\frac{1}{3}$ of 15.

So, $\frac{1}{3}$ of 15 is 5?

Yep.

How do we write that in math?

0   5   10   15

To find a fraction of a number, we **multiply** the fraction by the number.

$\frac{1}{3}$ **of** 15 means $\frac{1}{3}$ **times** 15.

$\frac{1}{3}$ of 15 means $\frac{1}{3} \times 15$

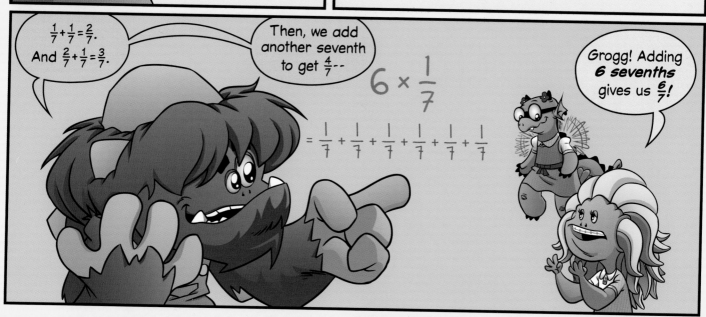

Oh, right. That makes sense.

6 times $\frac{1}{7}$ is 6 sevenths.

$$6 \times \frac{1}{7}$$

$$= \frac{1}{7} + \frac{1}{7} + \frac{1}{7} + \frac{1}{7} + \frac{1}{7} + \frac{1}{7}$$

$$= \frac{6}{7}$$

To find $6 \times \frac{1}{7}$ on the number line, we divide the number line into lengths of $\frac{1}{7}$ and count six of these lengths.

$$6 \times \frac{1}{7}$$

$$6 \times \frac{1}{7} = \frac{6}{7}.$$

Very good.

Let's try multiplying by a fraction whose numerator is not 1.

Who would like to try this one?

$$5 \times \frac{7}{8}$$

Try it.

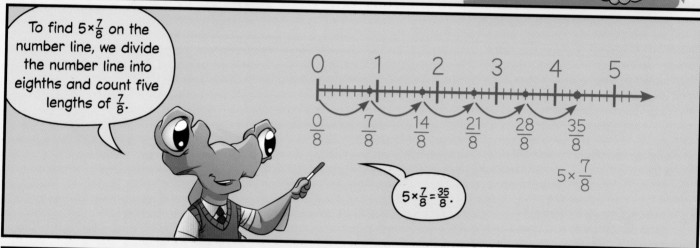

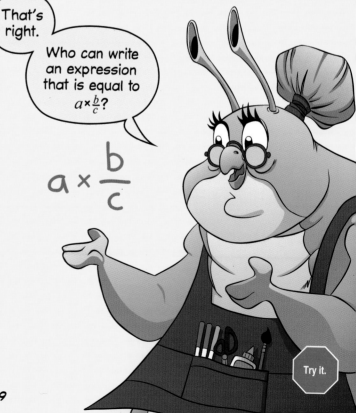

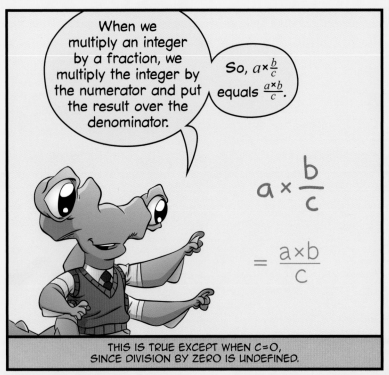

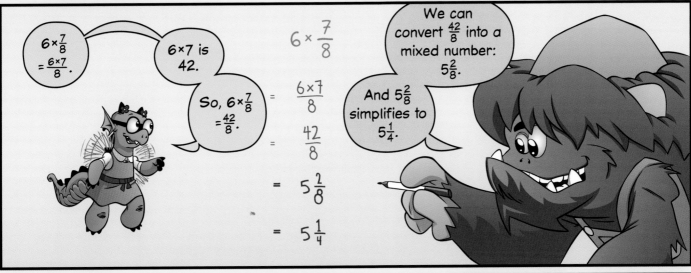

What number is $\frac{5}{8}$ of 3?

$\frac{5}{8}$ of 3

I know that $\frac{5}{8}$ is a little more than $\frac{1}{2}$.

So, $\frac{5}{8}$ of 3 is a little more than $\frac{1}{2}$ of 3.

$\frac{1}{2}$ of 3 is $1\frac{1}{2}$, so $\frac{5}{8}$ of 3 is a little more than $1\frac{1}{2}$.

Good estimating! Does anyone know how we could get an exact answer?

ESTIMATING THE ANSWER IS OFTEN A GREAT WAY TO BEGIN A PROBLEM.
YOU CAN ALSO USE YOUR ESTIMATE TO CHECK YOUR ANSWER WHEN YOU ARE DONE.

To find $\frac{5}{8}$ of 3, we multiply $\frac{5}{8} \times 3$.

$\frac{5}{8}$ of 3

$$= \frac{5}{8} \times 3$$

$$= \frac{5 \times 3}{8}$$

$$= \frac{15}{8}$$

$$= 1\frac{7}{8}$$

$\frac{5}{8} \times 3$ is $\frac{15}{8}$. We can write $\frac{15}{8}$ as a mixed number: $1\frac{7}{8}$.

That fits our estimate.

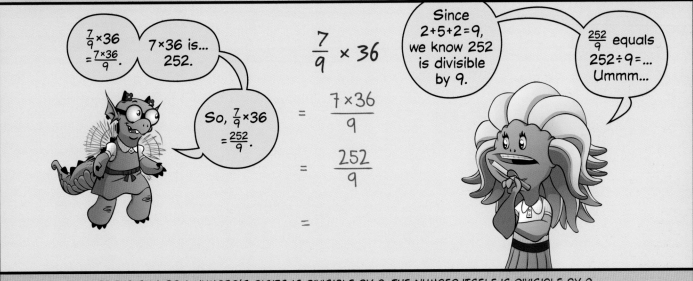

IF THE SUM OF A NUMBER'S DIGITS IS DIVISIBLE BY 9, THE NUMBER ITSELF IS DIVISIBLE BY 9. SEE BEAST ACADEMY 4C FOR AN EXPLANATION OF WHY THIS IS TRUE.

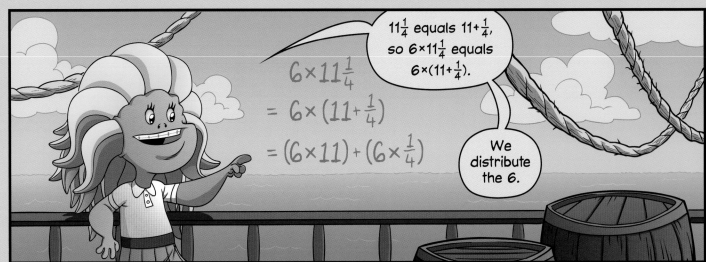

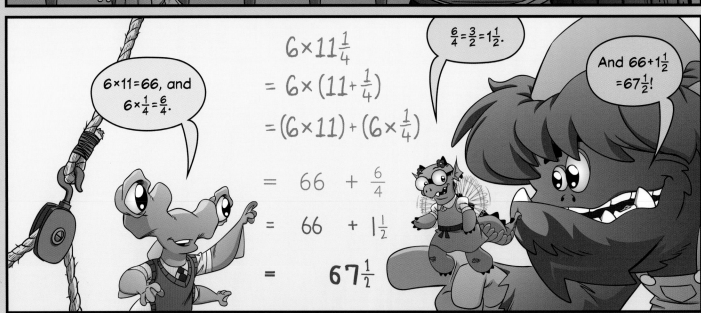

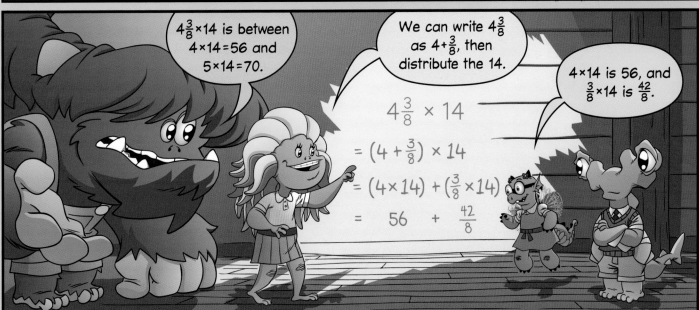

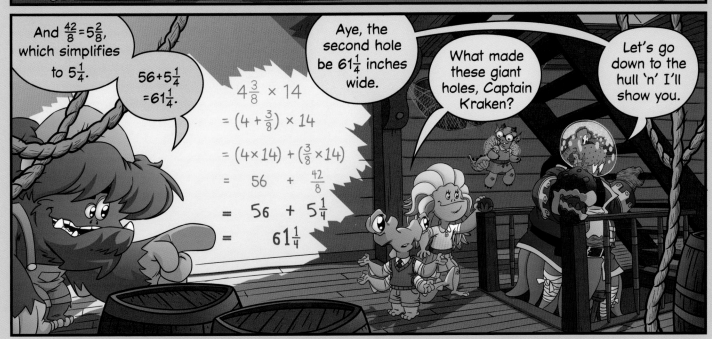

Homework: Write and solve a story problem involving fractions in the space below.

Contents: Chapter 11

See page 38 in the Practice book for a recommended reading/practice sequence for Chapter 11.

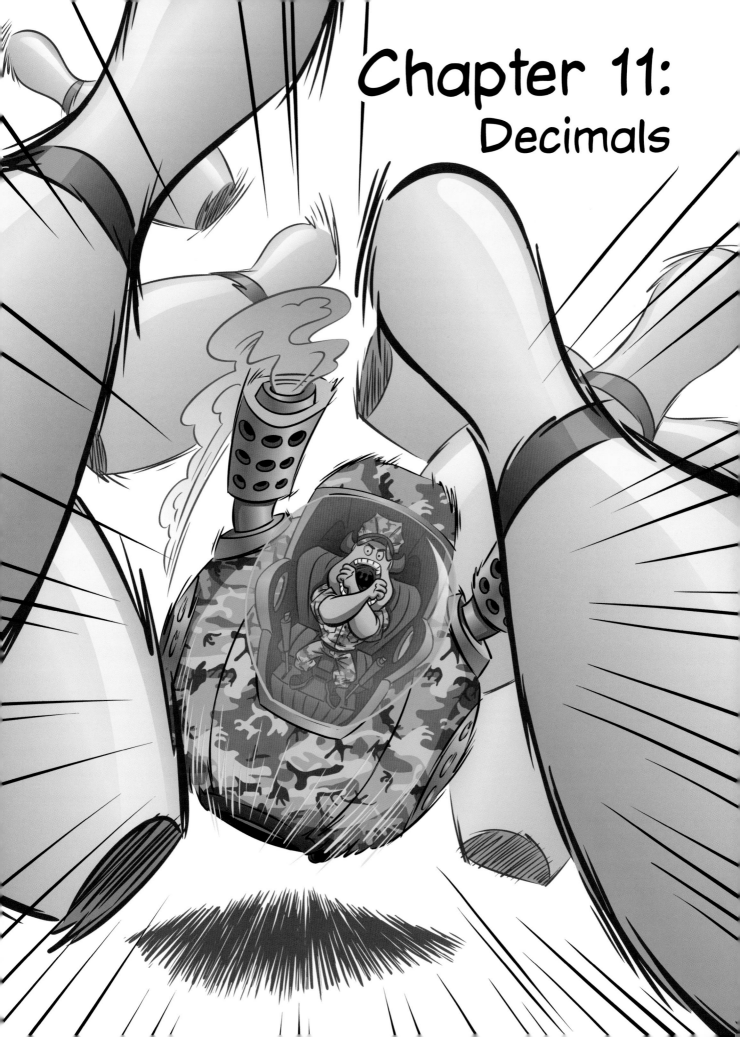

Chapter 11:
Decimals

7.3 IS READ "SEVEN POINT THREE." SIMILARLY, 12.2 IS READ "TWELVE POINT TWO," AND 2.05 IS READ "TWO POINT ZERO FIVE."

The dot is called a *decimal point.*

The decimal point separates the whole-number part from the fractional part.

Fractional part?

7.3

decimal point

The number 7.3 is a little more than 7.

To understand how much more, you need to understand *place value.*

I remember learning about place value!

Each place value is ten times the place value to its right.

That's right. But, how would you find the place values to the *right* of the ones place?

1,000,000's   100,000's   10,000's   1,000's   100's   10's   1's

×10   ×10   ×10   ×10   ×10   ×10

Try to fill in the place values to the right of the ones place.

41

We can simplify $12\frac{2}{10}$ to $12\frac{1}{5}$, and $2\frac{5}{100}$ to $2\frac{1}{20}$.

For the next couple of weeks before the Math Relays competition, we'll focus on decimal numbers.

In a decimal number, each place value to the right of the decimal point is a unit fraction whose denominator is a power of ten.

Place Values

$$100's \quad 10's \quad 1's \quad . \quad \frac{1}{10}'s \quad \frac{1}{100}'s \quad \frac{1}{1,000}'s \quad \frac{1}{10,000}'s$$

So, to write $\frac{7}{10}$ as a decimal number, we place a 7 in the tenths place.

$\frac{8}{10}$ can be written as 0.8.

And $\frac{9}{100}$ is written by placing a 9 in the hundredths place: 0.09.

$$\frac{7}{10} = 0.7$$

$$\frac{8}{10} = 0.8$$

$$\frac{9}{100} = 0.09$$

0.8 CAN ALSO BE WRITTEN WITHOUT THE ZERO, AS SIMPLY .8.
THE ZERO TO THE LEFT OF THE DECIMAL POINT MAKES THE NUMBER EASIER TO READ.

How do we write **ten** hundredths?

$$\frac{10}{100}$$

$\frac{10}{100}$ can be simplified to $\frac{1}{10}$.

So, $\frac{10}{100}$ can be written as $\frac{1}{10} = 0.1$.

$$\frac{10}{100} \overset{\div 10}{\underset{\div 10}{=}} \frac{1}{10} = 0.1$$

Oh, right.

How would we write $\frac{11}{100}$ as a decimal?

$$\frac{11}{100}$$

Try it.

DECIMAL NUMBERS ARE OFTEN SIMPLY CALLED "DECIMALS."

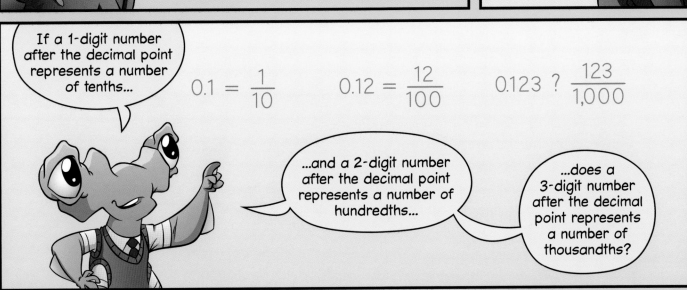
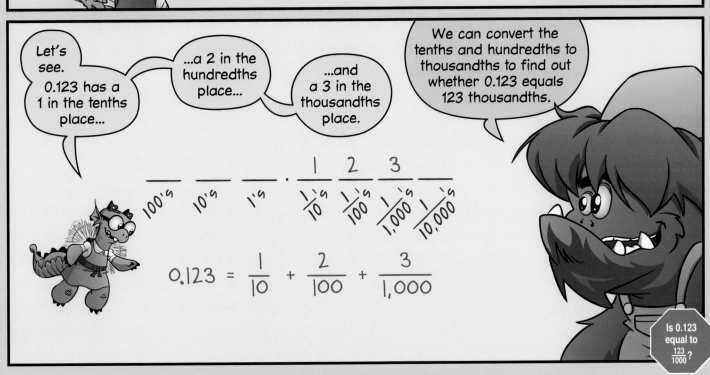

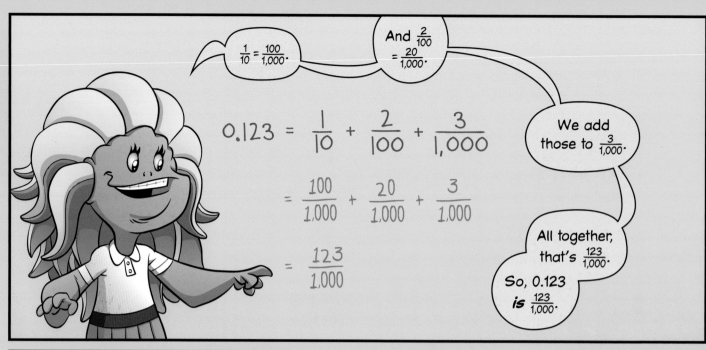

$$0.123 = \frac{1}{10} + \frac{2}{100} + \frac{3}{1,000}$$

$$= \frac{100}{1,000} + \frac{20}{1,000} + \frac{3}{1,000}$$

$$= \frac{123}{1,000}$$

$\frac{1}{10} = \frac{100}{1,000}.$

And $\frac{2}{100} = \frac{20}{1,000}.$

We add those to $\frac{3}{1,000}.$

All together, that's $\frac{123}{1,000}.$

So, 0.123 *is* $\frac{123}{1,000}.$

Nice job, little monsters.

A 3-digit number after the decimal point represents a number of thousandths.

$$0.817 = \frac{817}{1,000}$$

$$0.304 = \frac{304}{1,000}$$

$$0.019 = \frac{19}{1,000}$$

But...

...19 is a 2-digit number.

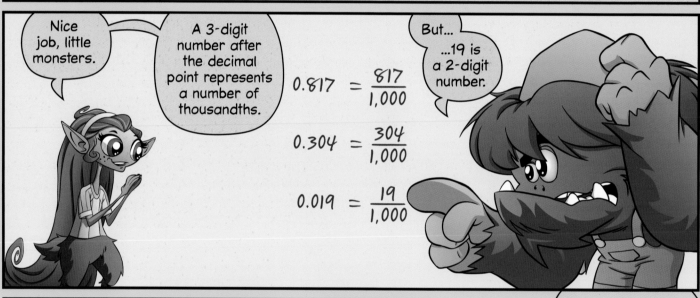

True, Grogg.

In a decimal number like 0.019, we have three digits to the right of the decimal point...

...but we don't write the zero in 019 when we write $\frac{19}{1,000}.$

Try writing each of these three decimals as a fraction or mixed number.

$$0.019 = \frac{0}{10} + \frac{1}{100} + \frac{9}{1,000}$$

$$= \frac{0}{1,000} + \frac{10}{1,000} + \frac{9}{1,000}$$

$$= \frac{19}{1,000}$$

0.059

0.006

3.0251

Try all three.

0.059 is 59 thousandths.

$$0.059 = \frac{59}{1,000}$$

0.006 is 6 thousandths.

$$0.006 = \frac{6}{1,000}$$

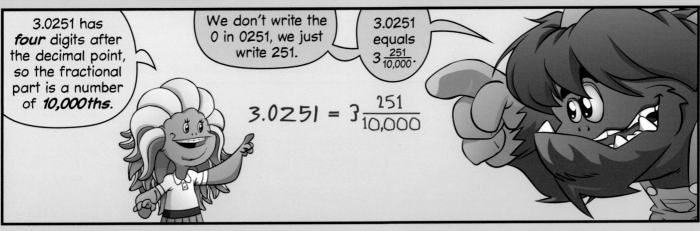

3.0251 has *four* digits after the decimal point, so the fractional part is a number of **10,000ths.**

We don't write the 0 in 0251, we just write 251.

3.0251 equals $3\frac{251}{10,000}$.

$$3.0251 = 3\frac{251}{10,000}$$

Maybe you could convince Ms. Q. to stop writing zeros in her grade book!

Last week I got two zeros for missing homework,

all due to an unfortunate incident involving a yeti and a frozen banana.

Sometimes it's good when Ms. Q. writes zeros in her grade book.

Huh?

You can't write 100 without two zeros.

49

Practice: Pages 39-49

Ms. Q.

Comparing Decimals

Which is greater: 1.7 or 1.08?

1.7   1.08

Hmmm...

Let's examine the facts.

1.08 has **more** digits...

...and 8 is **bigger** than 7.

Just because a number has more digits and the digits are bigger doesn't mean the number is bigger.

$\frac{888}{999}$ is **less** than $\frac{3}{2}$.

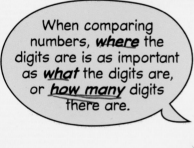

When comparing numbers, **where** the digits are is as important as **what** the digits are, or **how many** digits there are.

Since we know how to compare fractions...

...maybe we can write each decimal as a fraction, then compare.

Which is greater: 1.7 or 1.08?

50

When we put a 0 at the end of a positive integer, the other digits move to a larger place value.

_ 3 6
hundreds tens ones

3 6 0
hundreds tens ones

But, when we put a 0 at the end of a decimal, none of the other digits move.

So, when we add a 0 at the end of 1.7, we are just adding 0 hundredths.

1 . 7 _
ones tenths hundredths

1 . 7 0
ones tenths hundredths

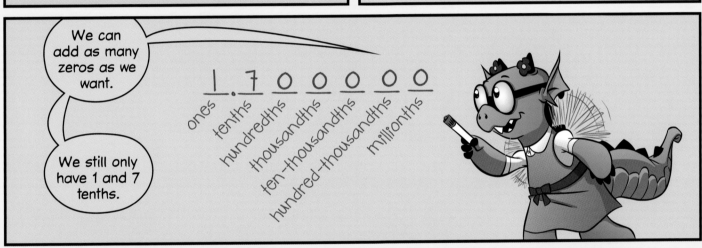

We can add as many zeros as we want.

We still only have 1 and 7 tenths.

1 . 7 0 0 0 0 0
ones tenths hundredths thousandths ten-thousandths hundred-thousandths millionths

Great explanation, Lizzie!

You can write zeros at the end of any decimal without changing its value.

How can this help you compare 7.275 to 7.3?

7.275   7.3

Which is greater: 7.275 or 7.3?

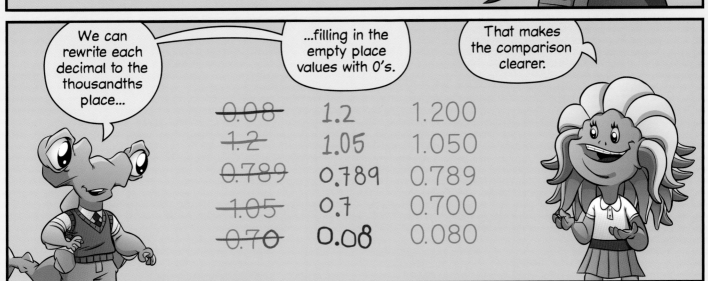

Name a decimal that is between 3 and 4.

Oh! I see...

I can write zeros as the hundredths digits of 3.1 and 3.2.

So, your number is between 3.10 and 3.20.

Yep.

krunch krunch

Krunch Krunch

3.14?

Close! Too low.

3.15?

Too high.

So, your number is between 3.14 and 3.15.

I think I know your number!

It's pi!

You got it!

THE NUMBER REPRESENTED BY THE GREEK LETTER PI (π) IS THE RESULT OF DIVIDING THE DISTANCE AROUND A CIRCLE BY THE DISTANCE ACROSS THE SAME CIRCLE. WHEN WRITTEN AS A DECIMAL, THE DIGITS OF PI GO ON AND ON FOREVER, STARTING WITH 3.14159265358979323846264338...

I get it!

*Point*less!

Did somebody say pie?

60

Practice: Pages 50-61

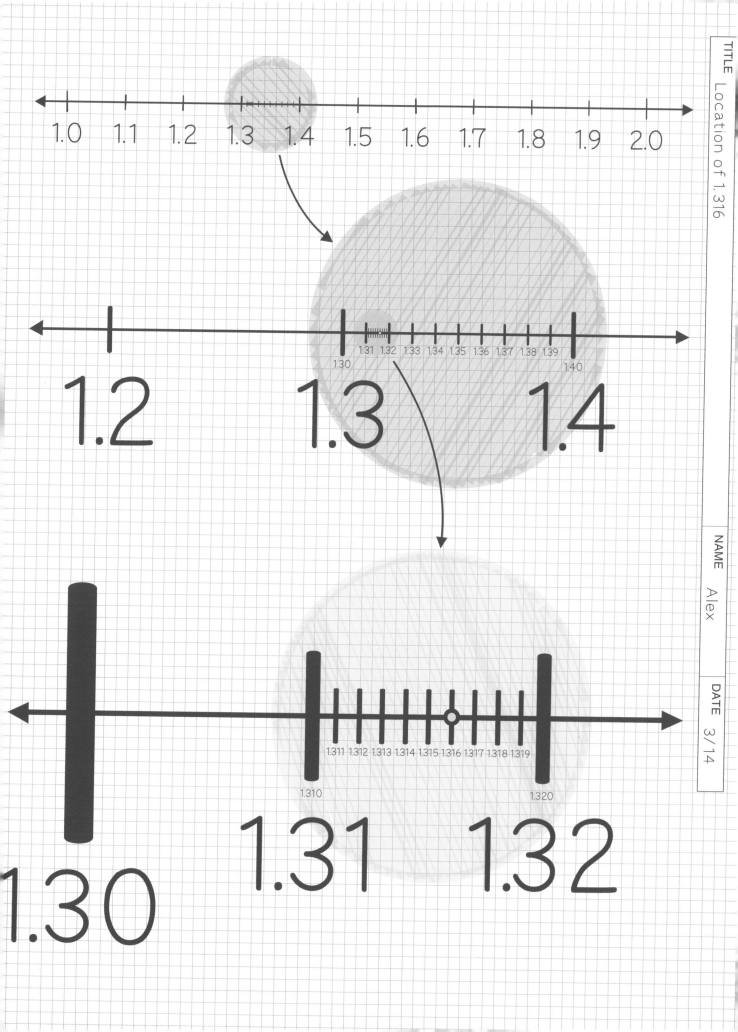

NAME Alex

DATE 3/14

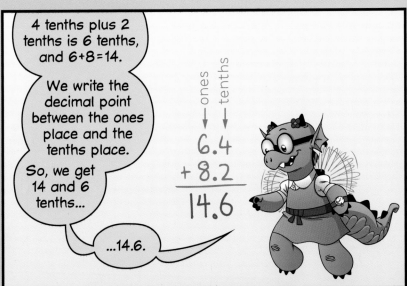

4 tenths plus 2 tenths is 6 tenths, and 6+8=14.

We write the decimal point between the ones place and the tenths place.

So, we get 14 and 6 tenths...

...14.6.

ones tenths

$$6.4$$
$$+8.2$$
$$\overline{14.6}$$

Very good. Try this one next.

What's 4.37 + 7.9?

4.37 + 7.9

Is this how we line up the numbers?

$$4.37$$
$$+7.9$$

The whole point of lining up the numbers is to align the place values.

We line up the *decimal points* of the numbers.

$$4.37$$
$$+7.9$$

That way, we add hundredths to hundredths...

...tenths to tenths...

...and the ones to the ones!

ones tenths hundredths

$$4.37$$
$$+7.90$$

We can write a zero at the end of 7.9 so that both numbers have the same number of digits to the right of the decimal point.

Add 4.37+7.9.

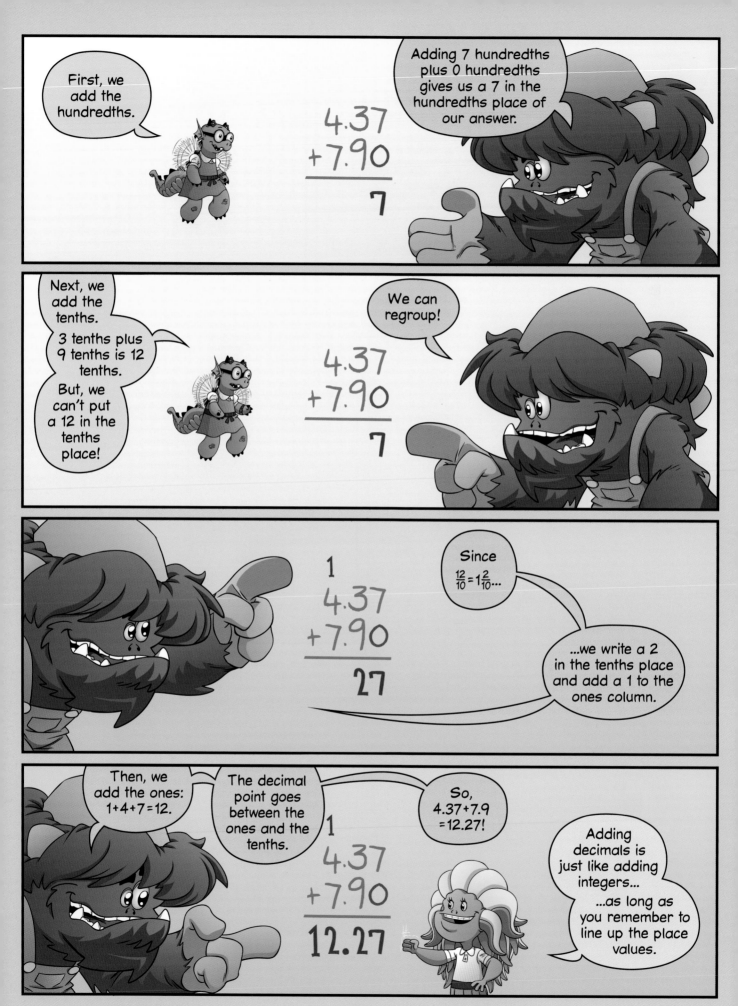

Subtraction doesn't seem any harder than addition.

I guess Calamitous Clod wasn't feeling very creative this week.

I'm pretty sure we just need to line up the place values...

...then subtract in the usual way.

39.42 – 6.803

```
  39.42
-  6.803
```

We start on the right with the thousandths.

But, there's nothing to subtract the 3 from.

```
  39.42
-  6.803
       -
```

We should write a zero at the end of 39.42.

That way, both numbers have the same number of digits to the right of the decimal point.

```
  39.420
-  6.803
```

Since we can't take away 3 thousandths from 0 thousandths...

...we need to take 1 hundredth from the hundredths place of 39.420 and break it into 10 thousandths.

That gives us a 1 in the hundredths place, and a 10 in the thousandths place.

```
        1  10
  39.4 2 Ø
-  6.8 0 3
```

Finish the computation.

68

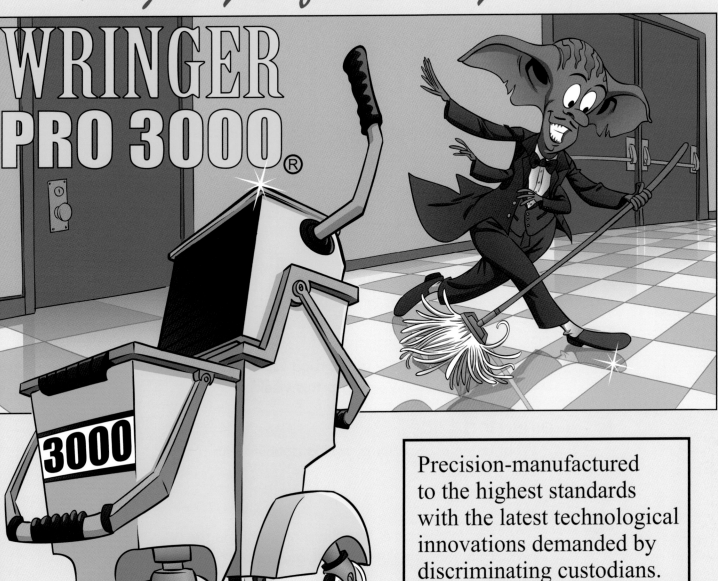

Contents: Chapter 12

See page 72 in the Practice book for a recommended reading/practice sequence for Chapter 12.

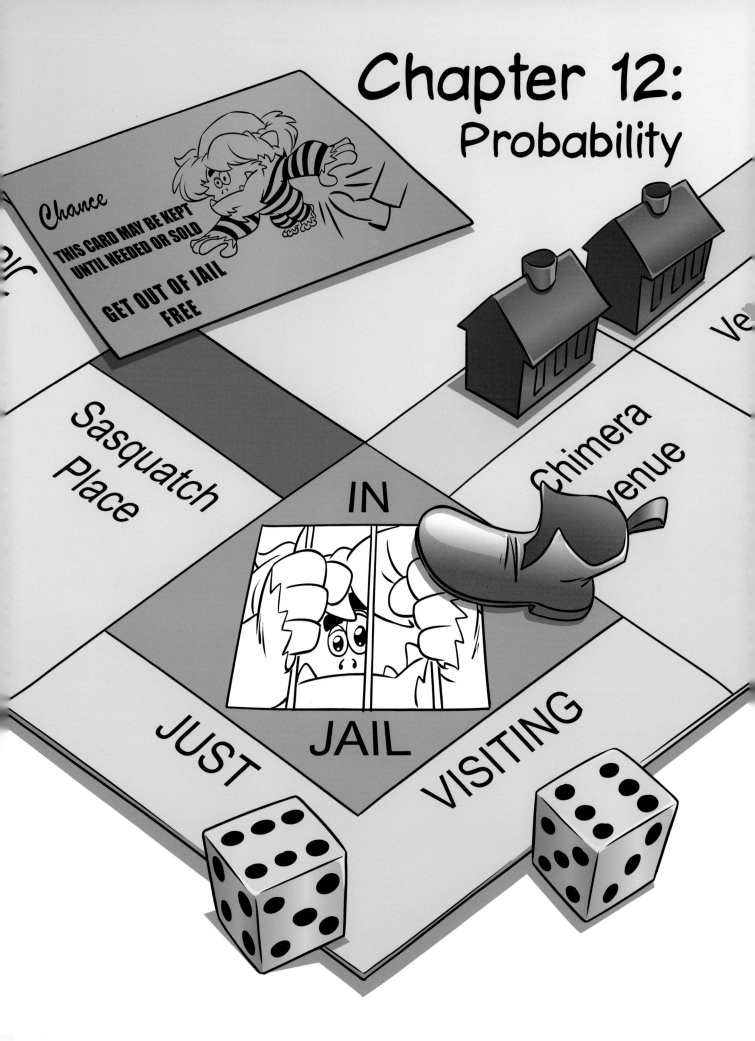

Chapter 12:
Probability

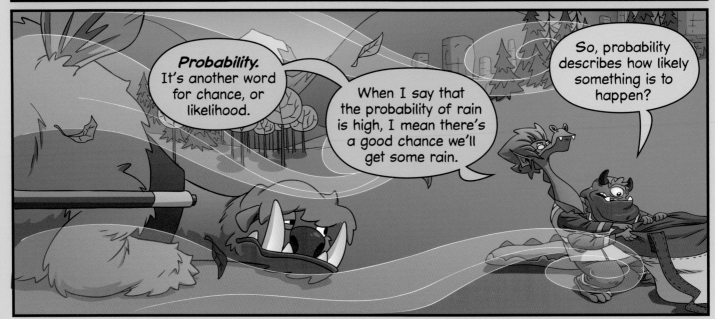

Exactly. If an event is likely, we say that it has a **high** probability of occurring.

That's right.

But having a **low** probability means an event is unlikely?

munch munch munch munch

It's already starting to drizzle. I think there are a few umbrellas on board.

Here they are. Which one should we use?

Let's leave it to chance.

Three of the umbrellas are yellow, and the fourth one is...

Ridiculous.

I'll turn off the lights and mix up all the umbrellas.

You reach in and grab the first umbrella you touch.

What kind of umbrella do you think you'll get?

What do you expect to happen?

73

74

Ms. Q.

Computing
Probability

Did everyone get a bag?

Yes!

Alex, what's in your bag?

Three red blocks and five yellow blocks...

...all the same size and shape.

If you pull a block from your bag without looking, what's the probability that the block will be yellow?

Hmmm... Since there are more yellow blocks than red blocks, I'd guess it's pretty likely.

How can we represent the probability of Alex drawing a yellow block from his bag?

An event's probability is usually expressed as a fraction.

We write the number of ways the desired event can happen over the total number of possible outcomes.

$$\frac{\text{Number of Desired Outcomes}}{\text{Number of Possible Outcomes}}$$

What is the probability that Alex will draw a yellow block?

THE WAYS A DESIRED EVENT CAN HAPPEN ARE CALLED *SUCCESSFUL OUTCOMES*, OR *FAVORABLE OUTCOMES*. THE NUMBER OF SUCCESSFUL OUTCOMES IS NEVER MORE THAN THE TOTAL NUMBER OF POSSIBLE OUTCOMES.

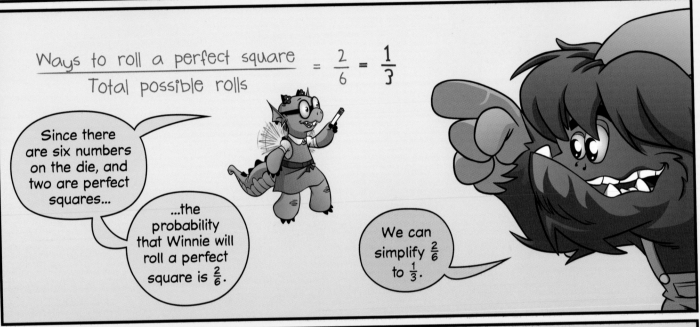

$$\frac{\text{Ways to roll a perfect square}}{\text{Total possible rolls}} = \frac{2}{6} = \frac{1}{3}$$

$\frac{2}{8}$ of the spinner is pink...

...so the probability of spinning pink is $\frac{2}{8}$, which equals $\frac{1}{4}$.

And the probability of spinning blue is also $\frac{1}{4}$.

Only $\frac{1}{8}$ of the spinner is green...

...so the probability of spinning green is $\frac{1}{8}$.

So, if you spun the spinner over and over again, you would get green about $\frac{1}{8}$ of the time, pink $\frac{1}{4}$ of the time, blue $\frac{1}{4}$ of the time...

...and purple about $\frac{3}{8}$ of the time.

Cool.

The spinner is 3 times more likely to land on purple than on green.

Probability of green:

Probability of purple:

$$\frac{1}{8} \quad \times 3 = \quad \frac{3}{8}$$

An event three times more likely than spinning purple would have a probability of $\frac{9}{8}$!

Grogg! Probability can never be greater than 1!

Huh?

Why can't the probability of an event ever be greater than 1?

Practice: Pages 73-81

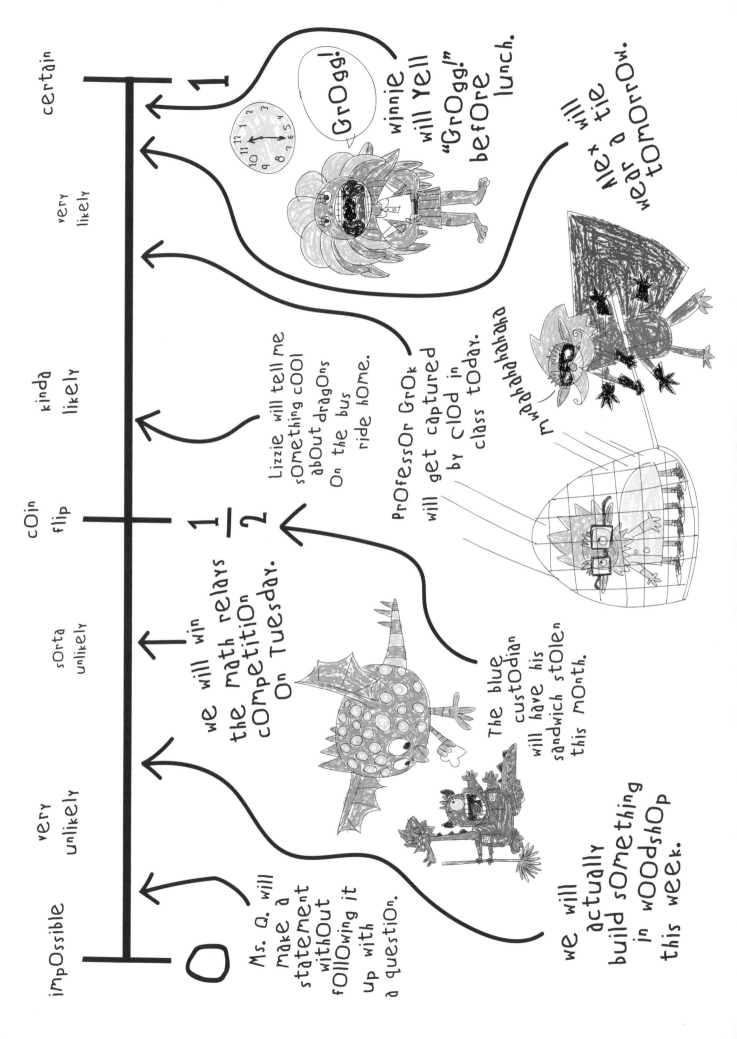

MATH TEAM
Coins and Dice

Alright, little monsters...

...climb aboard.

We're off to the math relays.

Do you think we have a chance of winning?

Some probabilities are almost impossible to compute.

But others aren't so tough.

For example, what is the probability that a tossed coin will land on heads?

It's a lot more likely to land on my head than on Lizzie's.

But, I guess it depends on where you aim it.

That is *not* what she means.

Most coins have a heads side, with the head of a famous monster...

...and a tails side, with the tail of the famous monster.

Landing heads means landing with the heads side face-up.

What is the probability of flipping heads on a coin flip?

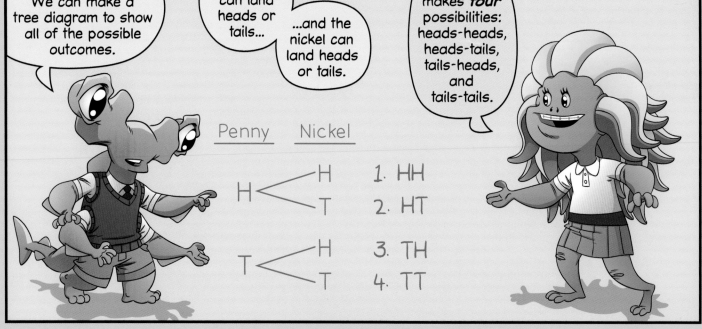

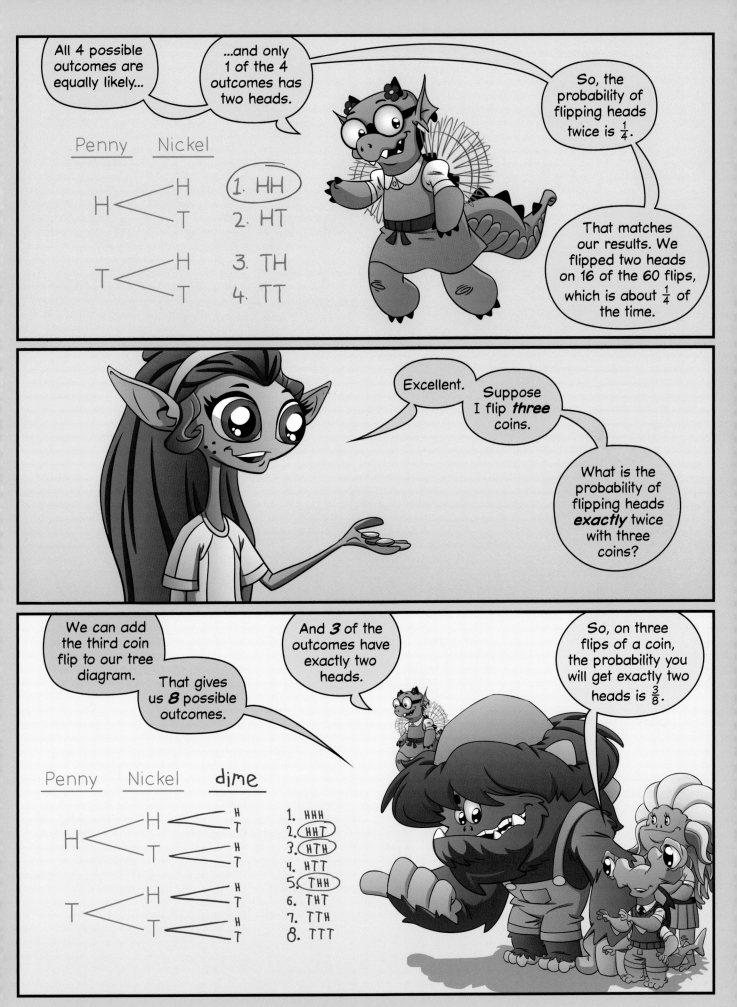

There are 6 possible rolls for the first die, and 6 possible rolls for the second die.

That makes 6×6=36 possible rolls.

1st Die Roll	2nd Die Roll	
1	1	1,1
	2	1,2
	3	1,3
	4	1,4
	5	1,5
	6	1,6
2	1	2,1
	2	2,2
	3	2,3
	4	2,4
	5	2,5
	6	2,6
3	1	3,1
	2	3,2
	3	3,3
	4	3,4
	5	3,5
	6	3,6
4	1	4,1
	2	4,2
	3	4,3
	4	4,4
	5	4,5
	6	4,6
5	1	5,1
	2	5,2
	3	5,3
	4	5,4
	5	5,5
	6	5,6
6	1	6,1
	2	6,2
	3	6,3
	4	6,4
	5	6,5
	6	6,6

There are 6 ways we could roll doubles, one for each number on the die.

6 of the 36 possible rolls are doubles, so the probability of rolling doubles is $\frac{6}{36}$...

...which simplifies to $\frac{1}{6}$.

$\frac{1}{6}$! That makes perfect sense!

What do you mean, Winnie?

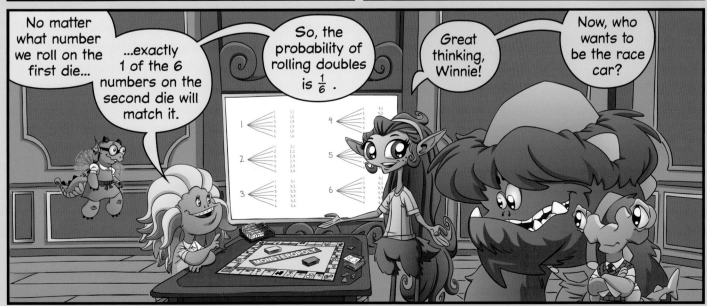

No matter what number we roll on the first die...

...exactly 1 of the 6 numbers on the second die will match it.

So, the probability of rolling doubles is $\frac{1}{6}$.

Great thinking, Winnie!

Now, who wants to be the race car?

7 heads in a row!?

Let me see that.

Heads. 8.

Amazing!

Huh?

I can't believe it landed heads again!

Why not?

The probability of flipping heads 8 times in a row is tiny.*

Yes, but each flip of the coin is as likely to land heads as tails.

*EACH FLIP HAS 2 POSSIBLE OUTCOMES. SO, FOR 8 FLIPS, THERE ARE $2 \times 2 \times 2 \times 2 \times 2 \times 2 \times 2 \times 2 = 256$ POSSIBLE OUTCOMES. ONLY ONE OF THESE OUTCOMES IS 8 HEADS IN A ROW, SO THE PROBABILITY OF FLIPPING 8 HEADS IN A ROW IS $\frac{1}{256}$.

So, it shouldn't be surprising each time it lands heads.

FLIP!

Catch

Heads.

How is that possible!?

You had already flipped heads 8 times in a row!

The coin didn't know that.

The previous tosses don't change the probability of the next toss landing heads.

But, the probability of flipping heads 9 times in a row is only $\frac{1}{512}$!

Sure, but the probability of flipping 8 heads followed by 1 tails is also $\frac{1}{512}$.

FOR 9 FLIPS, THERE ARE $2 \times 2 \times 2 \times 2 \times 2 \times 2 \times 2 \times 2 \times 2 = 512$ POSSIBLE OUTCOMES, ONLY ONE OF WHICH IS HHHHHHHHH, AND ONLY ONE OF WHICH IS HHHHHHHHT.

Ooooh. You're right.

Every sequence of 9 coin tosses is equally likely.

You are just as likely to toss 9 heads in a row as you are to toss H-H-T-H-T-T-T-H-H.

Right, but you'd probably never remember flipping H-H-T-H-T-T-T-H-H.

92

MATH TEAM
Math Relays: Round 1

We're almost there!

Before we land, let's review how the Math Relays competition works.

There are four problem stations in the competition hall.

I start at Station 1, Alex starts at Station 2, Winnie is at Station 3, and Grogg is at Station 4.

Other teams will have a player at each station, too.

When the relay begins, a problem appears on the screen at each station.

But the problems at stations 2, 3, and 4 all have blanks in them.

The correct answer to the problem at Station 1 fills the blank at Station 2.

I answer Problem 1 and bring it to Alex at Station 2.

We use the answer to fill in the blank in Problem 2, then we solve Problem 2 together.

I bring the Problem 2 answer to fill the blank at Station 3.

Alex and I work together to solve Problem 3.

Then I bring the Problem 3 answer to Station 4.

Winnie and I solve Problem 4...

...and I take the final answer to the Scorer's Table.

Four teams compete in each relay. The first two teams who deliver the correct answer from Station 4 to the scorer's table move on to the next round.

Look! We're about to land!

Here we are!

This is one of the biggest math events every year, so all of the top teams from the region are here.

There's Max and his team.

MATH RELAYS

He clobbered us all by himself at the Math Bowl.

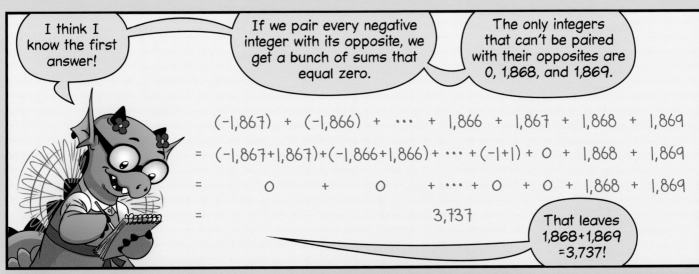

REVIEW UNITS–DIGIT COMPUTATIONS IN CHAPTER 3 OF GUIDE 4A.

I made a table to show the results of all 6×6=36 possible rolls.

There are 5 ways to roll a 6: 1+5, 2+4, 3+3, 4+2, and 5+1.

Since 5 of the 36 possible rolls give a total of 6...

...the probability of rolling a 6 is $\frac{5}{36}$.

		Second Die				
First Die	1	2	3	4	5	6
1	2	3	4	5	6	7
2	3	4	5	6	7	8
3	4	5	6	7	8	9
4	5	6	7	8	9	10
5	6	7	8	9	10	11
6	7	8	9	10	11	12

THIS TABLE GIVES ANOTHER WAY TO DISPLAY THE INFORMATION IN THE TREE DIAGRAM ON PAGE 89.

His teammate is already taking the final answer to the finish table!

Unbelievable. They finished before any of the other teams reached Station 3.

Let's head down to the arena floor. You guys compete soon.

As soon as your problem shows up on the screen, start working!

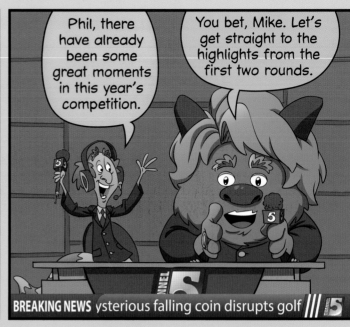

100

They have been an exciting team to watch, winning both their relays in impressive fashion.

Beast Academy hasn't sent a team to the Math Relays since their runner-up performance 6 years ago.

Back then, their current coach, Fiona Fawn, led the team to a memorable second-place finish.

COURTESY OF MATH RELAYS FILMS

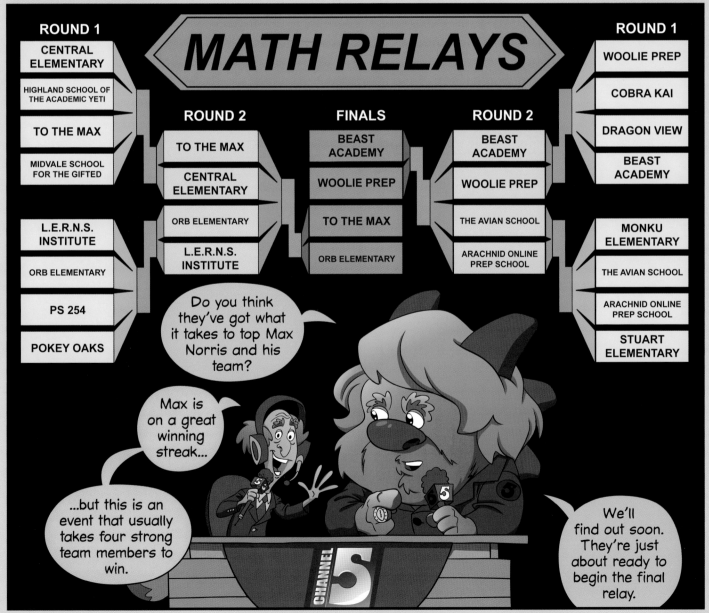

MATH RELAYS

ROUND 1

CENTRAL ELEMENTARY

HIGHLAND SCHOOL OF THE ACADEMIC YETI

TO THE MAX

MIDVALE SCHOOL FOR THE GIFTED

L.E.R.N.S. INSTITUTE

ORB ELEMENTARY

PS 254

POKEY OAKS

ROUND 2

TO THE MAX

CENTRAL ELEMENTARY

ORB ELEMENTARY

L.E.R.N.S. INSTITUTE

FINALS

BEAST ACADEMY

WOOLIE PREP

TO THE MAX

ORB ELEMENTARY

ROUND 2

BEAST ACADEMY

WOOLIE PREP

THE AVIAN SCHOOL

ARACHNID ONLINE PREP SCHOOL

ROUND 1

WOOLIE PREP

COBRA KAI

DRAGON VIEW

BEAST ACADEMY

MONKU ELEMENTARY

THE AVIAN SCHOOL

ARACHNID ONLINE PREP SCHOOL

STUART ELEMENTARY

Do you think they've got what it takes to top Max Norris and his team?

Max is on a great winning streak...

...but this is an event that usually takes four strong team members to win.

We'll find out soon. They're just about ready to begin the final relay.

This is what you've been working for. It's time to get out there and do your best.

Get to your stations.

Good luck!

ding ding

There's the start bell. The first four competitors have to solve Problem 1 without any help from teammates.

Problem 1:
One of the acute angles in a right triangle is 36 degrees larger than the other. How many degrees are in the measure of the smallest angle?

Beast Academy team member Lizzie is off to a quick start. Let's go to her glasses-cam for a better view.

It looks like she already has an answer!

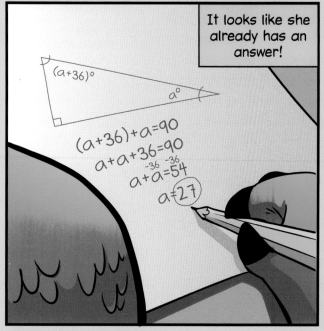

$(a+36)°$

$a°$

$(a+36)+a=90$

$a+a+36=90$

$\overset{-36}{}\overset{-36}{}$

$a+a=54$

$a=27$

Lizzie is headed to Station 2 just ahead of the other teams, Mike.

She'll join teammate Alex at Station 2.

Problem 2:
A fish tank that is $\frac{3}{4}$ *full* contains ____ gallons of water. How many gallons *must be* added to fill the tank?

Problem 3:
Find the sum of the five smallest multiples of ___ that have a remainder of 24 when divided by 25.

Meanwhile, the monsters at Station 3 are already hard at work.

And at Station 4, it looks like a few competitors are trying out some sample numbers, while Max's team member Dash Farnsworth appears to be stretching out his hamstrings.

Problem 4:
Two of the digits of ____ are selected at random and multiplied. What is the probability that their product is even?

Can you solve them all?

*THESE PAIRS ARE (2,7) (2,4) (2,5) (7,4) (7,5) AND (4,5). REVIEW COUNTING PAIRS IN BEAST ACADEMY 4B.

SOMETIMES IT'S EASIER TO COUNT THE POSSIBILITIES YOU *DON'T* WANT.

107

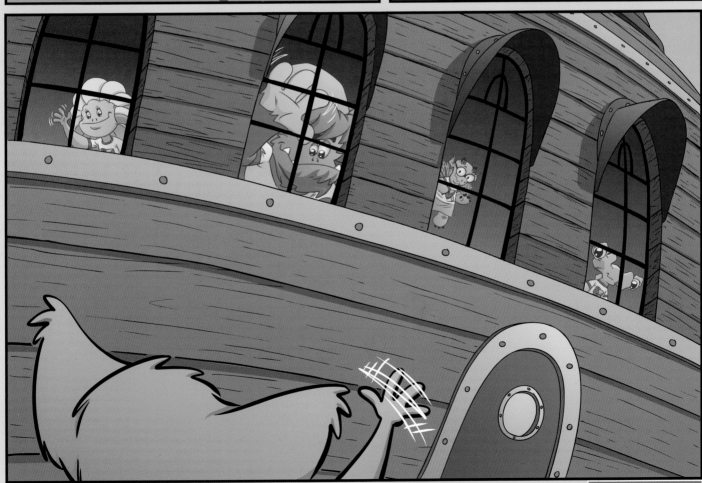

109

Practice: Pages 82-101

Index

Now Available!
Beast Academy Online

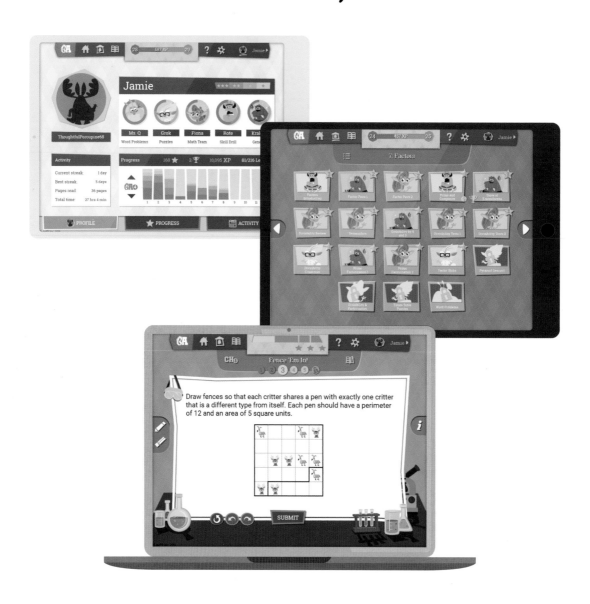

Learn more at BeastAcademy.com